Leaping Learners
Education, LLC

For more information and resources visit us at:

www.leapinglearnersed.com

Every attempt has been made to credit each photo. Please contact us if there has been an error and we will resolve the issue.

Cover, © soft_light/*stock.adobe.com*, Fish diver graphics © Alexey Bannykh/stock.adobe.com; pg.1 © *dynamofoto/stock.adobe.com, Pg. 10 © Susie/stock.adobe.com, pg11 © Fiona/stock.adobe.com, © p6m5/stock.adobe.com, pg 12 © paveu/stock.adobe.com, pg.17 © Andrea Izzotti/stock.adobe.com. pg. 20 © cherylvb/stock.adobe.com, pg.22 © davidevison/stock.adobe.com, pg 23 © Martin Wilkinson/stock.adobe.com, pg. 24 © aylerein/stock.adobe.com, pg 25 © Christos Georghiou, https://pixabay.com/en/stopwatch-racing-timer-timing-153398/, https://pixabay.com/en/scale-scales-cooking-food-kitchen-154924/*

All other photos done by Sean Bulger or royalty free from Pixabay.com
All design done by Sean Bulger

Special thanks to our reading level consultant
FRANCINE DEL VECCHIO, Ed.D.
Director of the Literacy Center
Professor of Education
Caldwell University

ISBN
978-1-948569-03-3

Dear Parents and Guardians,

Thank you for purchasing a *Fay Learns About* series book! After teaching students from kindergarten to second grade for over seven years, I became frustrated at the lack of engaging books that my students could read independently. To help my students engage with nonfiction topics, my wife and I decided to write non-fiction books for children. We hope to inspire young children to learn about the natural world.

Here at Leaping Learners Education, LLC, we have three main goals:

1. Spark young readers' curiosity about the natural world
2. Develop critical independent reading skills at an early age
3. Develop reading comprehension skills before and after reading

We hope that your child enjoys learning with Fay. If you or your children are interested in a topic that we have not written about yet, send us an email with your topic and maybe your book will be next!

Thank you,

Sean Bulger Ed.M

www.leapinglearnersed.com

Reading Suggestions:

Before reading this book, encourage your children to do a “picture walk,” where they skim through the book and look at the pictures to help them think about what they already know about the topic. Thinking about what they already know helps children get excited about learning more facts and begin reading with some confidence.

Preview any new vocabulary words with your child. Key vocabulary words are found on the last few pages of the book. Have your children use each new phrase in their own words to see if they understand the definition.

After previewing the book, encourage your child to read the book independently more than once. After they have read it, ask them specific questions related to the information in the book. Encourage them to go back and reread the relevent section in the book to retrieve the answer in case they forgot the facts.

Finally, see if your child can complete the reading comprehension exercises at the end of the book to strengthen their understanding of the topic!

This book is best for ages 7-8
but. . .
Please be mindful that reading levels are a guide and vary depending on a child’s skills and needs.

Fay Learns About . . . Sea turtles

Written by Sean and Anicia Bulger

Table of Contents

Introduction.....1

Body.....3

Habitat.....15

Diet.....19

Predators.....21

Babies.....23

Fun Facts.....27

Glossary.....28

Activities.....29

Hi! My name is Fay. I love to discover and learn new things. In this book, we will learn about sea turtles. Let's go!

Introduction

Sea turtles are both land and water **reptiles** that can live for a long time.

Reptiles

*Have scales

*Breathe air

*Lay eggs

Tricky word:
Chunk it out
Rep-tile

Body

Sea turtles have a head, four **flippers** and a tail.

They have a shell for protection.

Flipper
Shell
Tricky word:
Chunk it out
pro-tec-tion

Sea turtles have front flippers.

PG 6

Sea turtles have back flippers and a tail.

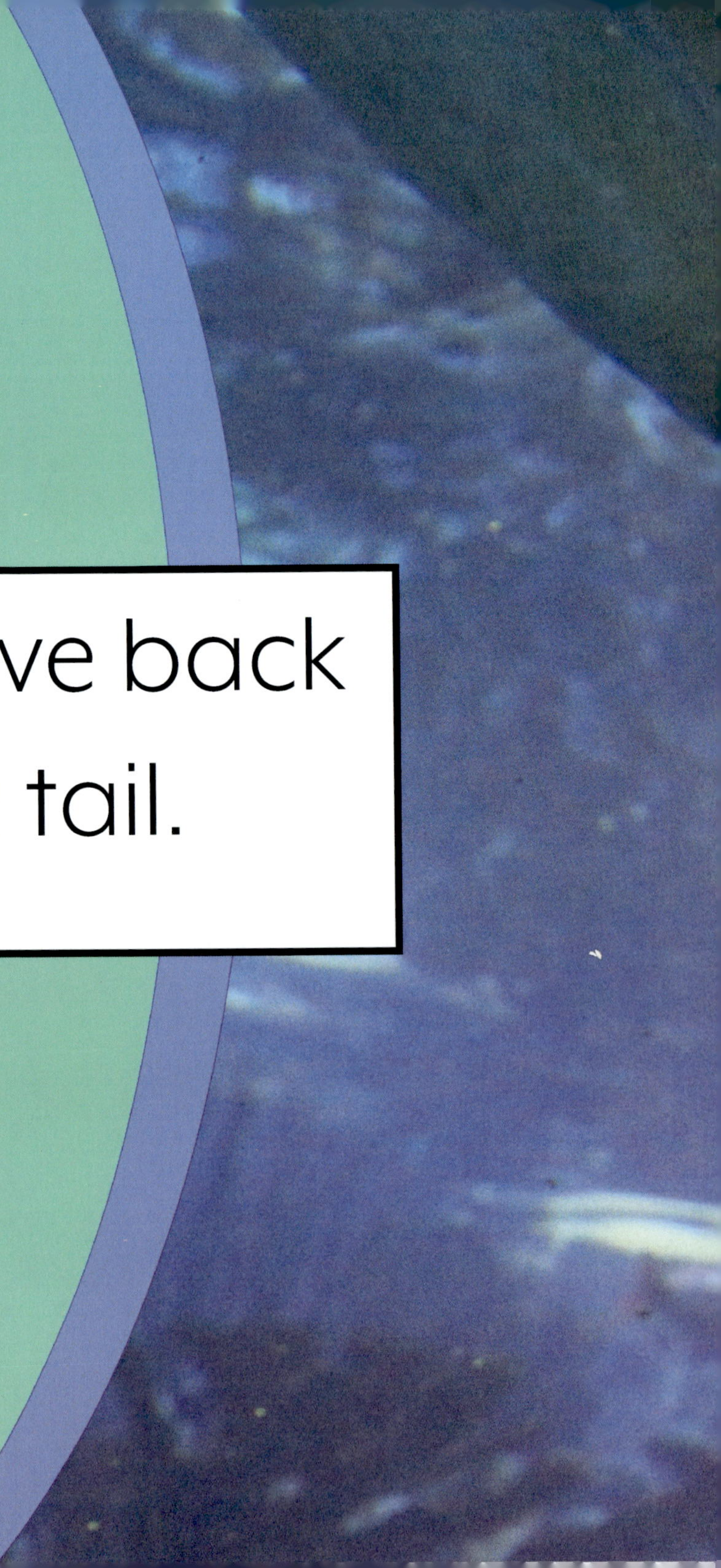

Tail
Flipper

Sea turtles use their flippers to move on land and in water.

Sea turtles have a stretchy neck to grab food.

Neck

Sea turtles use their **lungs** to breathe air.

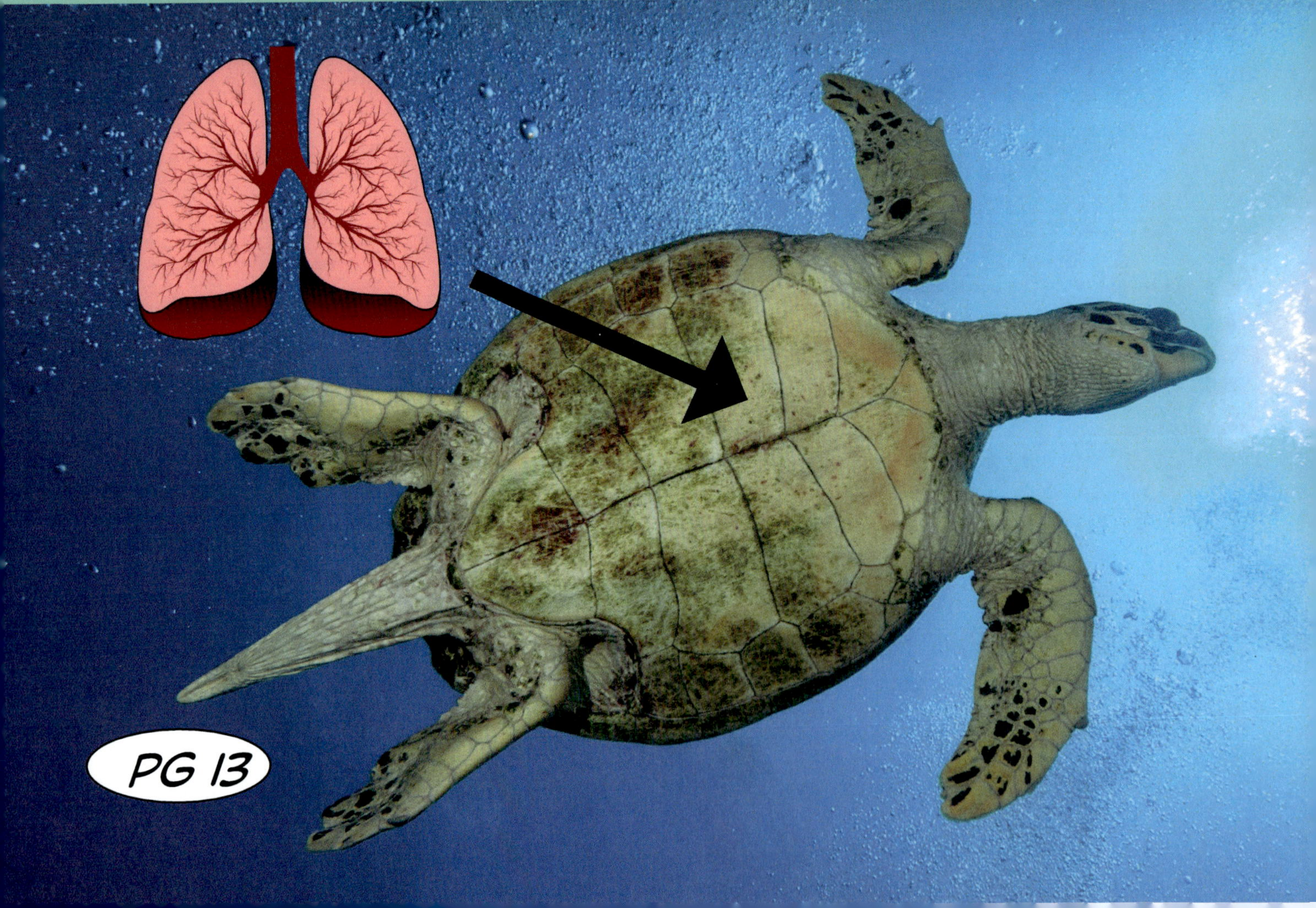

Sea turtles have lungs like people.
They do not have gills like fish.

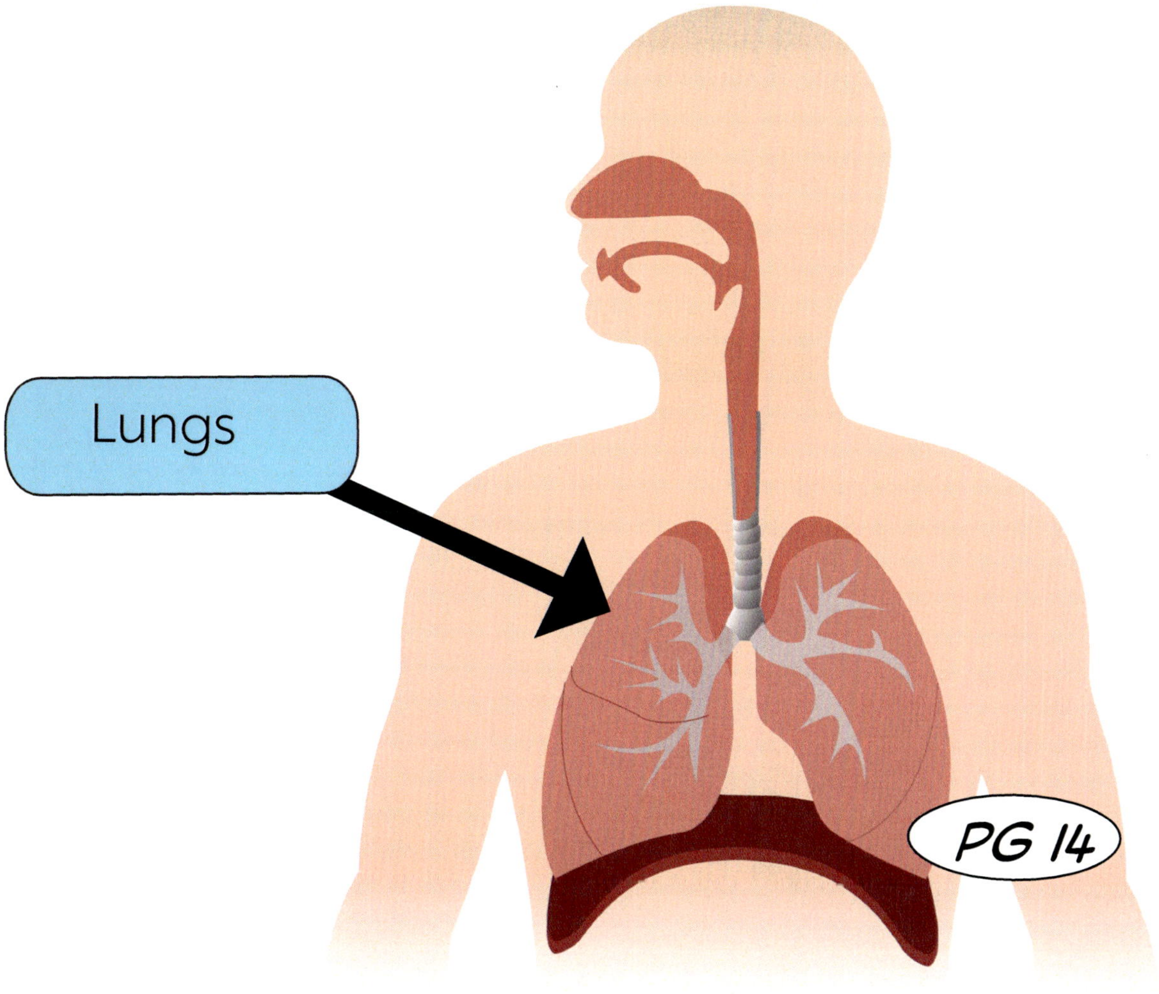

Habitat

Sea turtles live in warm oceans.

Sea turtles can also live on land.

Diet

Some sea turtles eat fish.

Some sea turtles eat plants.
Some eat plants and fish.

Predators

Not many animals hunt adult sea turtles. Their main **predators** are sharks.

Babies

Sea turtles bury their eggs in the sand and leave them there to **hatch**.

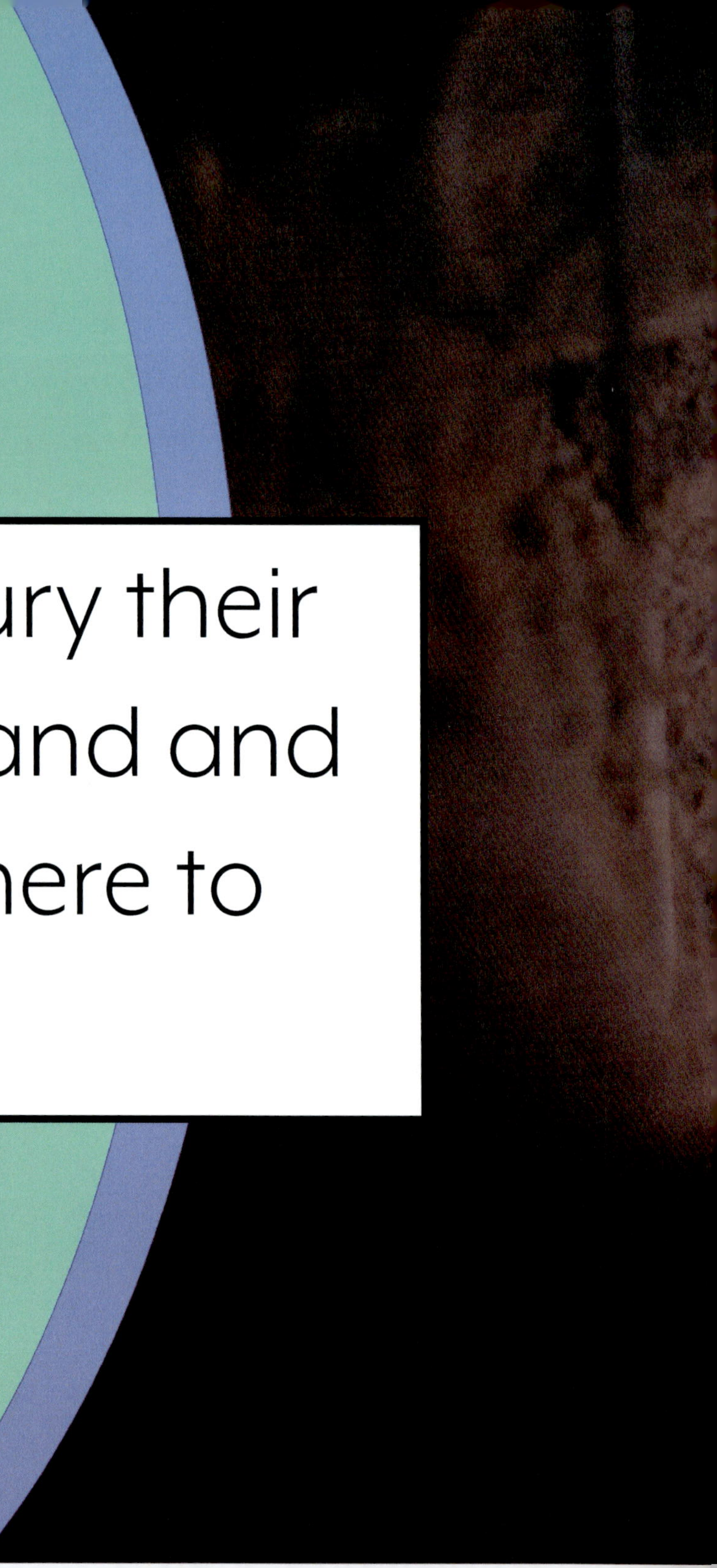

Eggs

Baby sea turtles are small.

Baby sea turtles are born in large groups. They rush to the ocean after they hatch to escape predators.

The oldest known sea turtle is 400 years old.

The leatherback sea turtle can weigh up to 2,000 pounds.

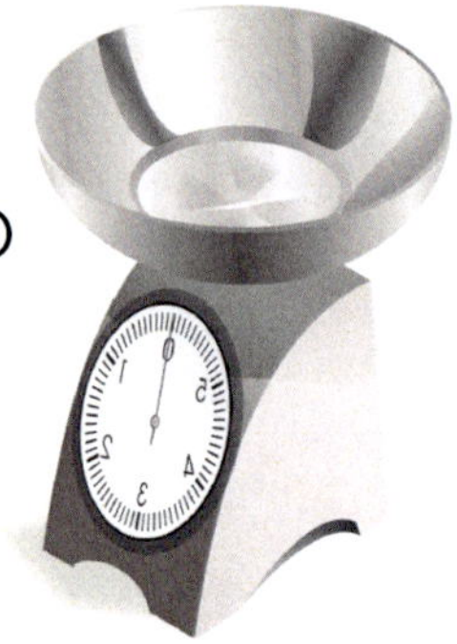

A sea turtle can hold its breath for 4-7 hours.

Glossary

A glossary tells the reader the meaning of important words.

Reptile – A cold-blooded animal that lays eggs and has scales

Flipper – Flat body part that helps an animal swim in water

Lung – An organ animals use to breathe air

Predator – Animal that eats other animals

Hatch – When a baby is born out of an egg

PG 28

Draw a picture of a sea turtle

Choose 3 words from the glossary and write a sentence for each one.

1.______________________________

____

____

2.______________________________

____

____

3.______________________________

____

____

Quiz

1. What helps sea turtles move?

a. Suction cups

b. Flippers

c. Feet

2. What animals eat sea turtles?

a. Eels

b. Squid

c. Sharks

3. How does a shell help sea turtles?

a. Gives it energy

b. Protects it

c. Attracts other sea turtles

Common core standards:

RI. 2. 1 - Questions 1, 2, 3

4. Which section did you learn about a sea turtle's neck?

a. Body

b. Diet

c. Habitat

5. How many flippers do sea turtles have?

a. 2

b. 4

c. 8

6. How long can sea turtles hold their breath?

a. 4-7 min

b. 4-7 hours

c. 1 hour

Common core standards:
RI. 2. 1 - Questions 5,6
RI. 2. 2 - Question 4

Answer the following questions using information from the text:

1. What do sea turtles eat?

____

____

____

____

____

2. How do sea turtles stay safe?

____

____

____

____

____

____

3. How do sea turtles breathe?

4. Where do sea turtles lay their eggs?

Want to learn about rainforest animals? Check out the "Matt Learns About..." Series!

Great for emerging readers ages 6-8

Want to learn about colors? Check out the "Clayton Teaches you About..." Series!

Great for early readers ages 4-6

Want to learn about Farm Animals? Check out the "Katie Teaches you About..." Series!

Great for early readers ages 4-6